ARTIFICIAL INTELLIGENCE

THE FUNDAMENTALS OF ARTIFICIAL INTELLIGENCE AND MACHINE LEARNING

TIM D. WASHINGTON

Copyright 2019 by Content Arcade Publishing-All rights reserved.

This document is geared towards providing exact and reliable information in regards to the topic and issue covered. The publication is sold with the idea that the publisher is not required to render an accounting, officially permitted, or otherwise, qualified services. If advice is necessary, legal or professional, a practiced individual in the profession should be ordered.

From a Declaration of Principles which was accepted and approved equally by a Committee of the American Bar Association and a Committee of Publishers and Associations.

In no way is it legal to reproduce, duplicate, or transmit any part of this document in either electronic means or printed format. Recording of this publication is strictly prohibited, and any storage of this document is not allowed unless with written permission from the publisher. All rights reserved.

The information provided herein is stated to be truthful and consistent, in that any liability, in terms of inattention or otherwise, by any usage or abuse of any policies, processes, or directions contained within is the solitary and utter responsibility of the recipient reader. Under no circumstances will any legal responsibility or blame be held against the publisher for any reparation, damages, or monetary loss due to the information herein, either directly or indirectly.

Respective authors own all copyrights not held by the publisher.

The information herein is offered for informational purposes solely and is universal as so. The presentation of the information is without a contract or any type of guarantee.

The trademarks that are used are without any consent, and the publication of the trademark is without permission or backing by the trademark owner. All trademarks and brands within this book are for clarifying purposes only and are owned by the owners themselves, not affiliated with this document.

Table of Contents

INTRODUCTION i
Chapter 1: ARTIFICIAL BEINGS, A BRIEF HISTORY OF THE HUMAN PSYCHE 1
Chapter 2: TOP SIX AI MYTHS 6
Chapter 3: WHY AI IS THE NEW BUSINESS DEGREE 11
Chapter 4: UNDERSTANDING MACHINE LEARNING 16
    What Is Machine Learning? 16
    How Machine Learning Differs From Artificial Intelligence 17
    How Machine Learning Works 17
Chapter 5: MACHINE LEARNING STEPS 18
    Real Life Problem 18
    Step 1: Creation of a model 19
    Step 2: Providing input 19
    Step 3: Learning 20
    Unsupervised Machine Learning 21
    Reinforcement Machine Learning 21
    Limitations of Machine Learning 22
        Time constraints 23
        Imperfect prediction 23
        Variability problem 23
        Uses of machine algorithm is limited 24
Chapter 6: ROBOTICS 25
    Artificial Intelligence and Robots 28
Chapter 7: NATURAL LANGUAGE PROCESSING 30
    What is it and What is it Used For? 30
    Well, Why NLP? 31
    Well, How Does NLP Work? 32
    Semantic Analysis 32
    Problems 33
    Activities Involved in Natural Language Processing 33
    How NLP is Changing Search and Customer Service 35
CONCLUSION 37

INTRODUCTION

Artificial Intelligence is a field that has a long history but has continued to grow and change. It is a powerful driving force in changing humanity by assisting businesses and people generate exciting, creative products and services, make critical decisions, and attain major goals. This is the reason why organizations keep hiring AI experts at a jaw-dropping speed.

The median salary of an AI developer in the US is not less than $80,000 based on payscale.com. Virtually all great tech companies run an artificial intelligence project and are ready to cash out millions of dollars to assist in completing the project.

Approximately 13.6 million jobs will emerge in the AI field in the next decade. However, there is a staggering shortage of talent in AI. For example, there are less than 10,000 people in the world with the skill set sufficient to carry out a significant research.

Artificial Intelligence (AI) technology is highly popular in our daily lives. It has applications in different sectors right from gaming, media to finance, and also the state-of-the-art research industries from medical diagnosis, robotics, and quantum science. The Artificial Intelligence Book for beginners focuses on driving an interest in its learners in the field of AI so that they are ready to learn more advanced topics in the same field.

The introduction to AI explains the background history of artificial intelligence, robotics, learning methods of Artificial Intelligence, basics of modern AI, and some representative of applications of AI. Along the way, we also plan to excite you about the huge possibilities in the field of AI, which continues to drive human ability beyond our imagination.

Chapter 1: ARTIFICIAL BEINGS, A BRIEF HISTORY OF THE HUMAN PSYCHE

Believe it or not, artificial intelligence dates back to antiquity, long before computers were even invented. The first mentions of artificial agents can be traced to Greeks myths like the tale of a giant bronze automaton Talos, tasked with the protection of Crete from invaders. Talos was defeated by the sorceress Medea when she removed a bronze nail keeping in a type of liquid or lifeblood, possibly fuel. Another myth tells the story of a sculptor, Pygmalion, who creates a statue of a beautiful woman, only to witness it come to life before his eyes. These stories are perhaps the first mentions of the robot trope in recorded history. The creation of these artificial beings has been a reoccurring human fascination since then. They can be observed in Greek and Arabic literature. In medieval times, the Swiss alchemist Paracelsus claimed to have created a homunculus or artificial being with nothing more than his sperm, magnetism, and alchemy. The Jewish rabbi Maharal of Prague is associated with a legend of the clay golem he created to defend Jews from persecution.

It was perhaps this fascination with the artificial that led tinkerers to create elaborate mechanical sculptures or automatons that moved into place. Though back then people didn't have the computing power to simulate intelligence, they could still use mechanics to simulate motion. More elaborate automata, like the ones designed by Ismail al-Jazari, moved away from pure mechanics and used hydropower. Some of his creations included a peacock fountain that served as a hand washing station. A secret compartment would even offer a bar of soap with the movement of the peacock. After the medieval age passed, automatons persisted into the coming centuries. Some of the theories of mind posited by the philosopher Rene Descartes stemmed from a visit he made to an automata garden at Saint-Germain-en-Laye, Paris. Descartes observed that if an automaton could be motivated to move by the flow of water, a human could be motivated by the existence of the mind as a substance. Another word for this is the soul, or later as the "ghost in the machine". He viewed the body as a purely mechanical vessel that was driven by the mind, an immaterial substance distinct from even the brain. The 18th century also saw the creation of the infamous Turk, an automaton whose inventor claimed to be able to play chess on its own. It was later revealed to be a hoax, operated by a human, but it was nevertheless intriguing. To think that even in the 18th century, people were going about creating systems to play chess against human players! This was long before IBM's Deep Blue beat the world chess champion Garry Kasparov in 1996 and long before AlphaGoZero defeated the Go champion. Amazon's crowd intelligence platform Mechanical Turk is fittingly named after the automaton.

Interest with mechanical behavior in the 19th century was embodied in the work of E.T.A Hoffman, a writer known for creating a feeling of unease in his stories. More specifically, the psychologist Sigmund Freud used the term **Unheimlich** or "uncanny" to describe the feeling he felt from reading Hoffman's writing. The story that is given the most attention is **Der Sandman** or **The Sandman** in English. It's a story of a mysterious figure from folklore named the Sandman that steals the eyeballs of young children to feed his own offspring. A character in the story named Olympia is introduced as the daughter of a professor but later revealed to be an automaton – a doll of his own creation. Olympia is striking because she is virtually indistinguishable from a young woman in the story, so much so that the protagonist falls in love with her and proposes marriage. However, just as he is about to propose, he comes across the professor fighting over the doll's lifeless body with his collaborator, arguing over who designed the eyelids and who made the clockwork mechanisms that power her. The protagonist sees Olympia's glass eyeballs strewn on the floor and goes mad. The automaton is essentially a mannequin with all the likeness of a real person – a cruel experiment carried out by the professor and his collaborator.

The same concept of uncanniness led roboticist Masahiro Mori to coin the term "uncanny valley" to describe the emotional response people have to lifelike robots. In general, the more photorealistic a robot is, the more uneasy people feel. It's a strange feeling, like seeing a real caricature of life right in front of you, the distinction between real and artificial completely blurred. However, the viewer nevertheless understands that the thing they see before them is fake. The robotic figure then exudes a cold, impersonal atmosphere that makes the hairs of the back stand up. The uncanny valley refers to a graphical chart with human likeness on the x-axis and familiarity on the y-axis. As human likeness of a robot increases, familiarity drops indicated by a distinct downward curve. At the lowest point of the curve are totally realistic depictions of people that are lifeless like corpses and zombies. These things elicit an uncanny effect because they are familiar, but we know that they are unliving. There is a strangeness associated with that loss of consciousness that is inadvertently reproduced in a robot that looks like a human, but that is also not living. Give this realistic robot a voice and some semblance of intelligence, and you get a creepy aberration of the real thing.

Though normally applied to robots, it is easy to see how the uncanny valley can apply to pure artificial intelligence or depictions of it. HAL 9000 is a computer, yet it creates a feeling of uncanniness when its logical response to an illogical situation results in a feeling of dread. If you saw the movie in theaters when it first came out, you could have heard a loud gasp ring out from the audience when they realized that HAL was reading the lips of the crew members talking about disconnecting it.

The imperfection of artificial intelligence and the breakdown or malfunction of such systems was characterized in Herman Melville's **Bartleby the Scrivener**. Bartleby, a Wall Street clerk, has a sudden mental breakdown in the middle of his work, merely proclaiming, "I would prefer not to" when asked to perform a task. The character repeats this signature phrase over and over again to the point that it becomes a robotic drawl. The character exhibits other robotic qualities as well, like staring off into space at a brick wall, as if waiting for input.

Another popular work that was published around the same time was Marry Shelley's **Frankenstein** in 1818. The original subtitle of the book was "Or The Modern Day Prometheus", but this has been dropped in most recent publications of the book. In this timeless classic, Shelley explores the ethics of creating artificial beings, how they may act, and what humans can expect. The story takes on a humanitarian perspective, as Frankenstein's monster develops feelings of alienation after realizing that he is of a different kind. After his creator rejects to create a female version of himself, the creature murders his fiancée, making things even. The creature laments that as a living being with sentience, he has the right to happiness, a right that his creator has deprived him off.

This early interpretation of AI, Frankenstein is wary of the two artificial beings breeding and creating an unstoppable race that subjugates humanity under their evil. However, besides the killing of the fiancée as revenge, the creature is never patently evil. It is only because the creature has a menacing appearance that he is branded as such. The fears of the two creatures breeding are consistent with the fear of an intelligence explosion, as any sufficiently intelligent artificial being could create clones of itself if it so wished.

It is the mechanical and computational aspects of artificial intelligence that scares people. It is hard to say whether modern doomsday scenarios rooted in AI come from this fear or if they stem from the recent advances in machine learning. Whatever the case, it's nothing too new. Humanity is still continuing its never-ending quest to create artificial minds as it always has been since the very beginning. But even with all the computing power available in the world, we still do not know how these artificial minds will come about.

Chapter 2:
TOP SIX AI MYTHS

Given the increasing frequency of AI being talked about in popular news media as well as in academic sources, it is difficult for a novice to separate what is fact and fiction. They are also likely to form their own conclusions about the nature of AI without first doing their research. This is in one sense dangerous because sensationalist headlines can morph public opinion, sometimes without a person having even read the article. It is in another sense doing a disservice to the consumer, the student, the voter, and the interested layperson because they may form faulty or misinformed conclusions about AI. While AI is a complex field with a rich history, it doesn't take an expert or a historian to approach AI with a critical eye. The truth is – there are many people making publications on AI, commenting and forming predictions, who at the same time have zero formal training in it. For these reasons, this chapter is dedicated to the most common AI myths perpetuated by mass media, folklore, and popular opinion. Some of these myths have already been covered in previous chapters or in various degrees of scrutiny, but they will be laid out here to bring the message home in case you missed their importance. Some of them will also be touched upon later in the book.

Myth #1: Machine learning is the same thing as AI

The focus on machine learning algorithms make it seem like "machine learning" is the same thing as artificial intelligence. Machine learning gets all the attention because it is "sexy" and currently the biggest area of research. Artificial intelligence is the study and engineering discipline of programming computers to perform tasks previously thought required human intelligence.

When Deep Blue defeated the chess champion in the 1990s, it wasn't using machine learning. Deep Blue was simply a really fast computer that could predict the best moves based off computing all possible future moves. It was, in essence, a brute force approach to defeating a world-class chess player. Kasparov had to rely on the information processing of his mind alone, and so he lost.

Myth #2: Machine Learning is how computers will learn how to become smart

Nobody actually knows how general intelligence will come about. When people hear about machine learning, their first thought is that researchers are teaching computers how to be smart. In reality, they are only training algorithms to perform tasks accurately. Researchers first used artificial neural networks because they believed that logical abstractions were enough to simulate intelligence. And they mostly succeeded. Though machine learning and neural nets have their limits, they still do a fine job at what they were designed to do.

Myth #3: AI can understand language

This is the same point that John Searle was trying to get across. The difficulty in imagining how artificial general intelligence will work is the same difficulty in imagining how a computer might understand something as complex as language. At a fundamental level, a computer only understands logical constructs. Ones, zeroes, and logic gates are all that they operate on. Language has traditionally been a difficult area to tackle with machine learning. The failure of machine learning to translate human language even after years of research was one of the catalysts for AI winter. Some said it simply could not be done. Now, we have machine translation algorithms like the ones used by Google Translate. These are still imperfect as anyone who used them can attest, but they are a step in the right direction. Teaching a machine to parse language belongs to an interdisciplinary field called natural language processing (NLP). It is an intersection of linguistics, computer science, psychology, and artificial intelligence. However, even with NLP, the computer is still just doing a bunch of fancy algorithms. It doesn't intuitively understand language at all.

Myth #4: AI programs can modify their own code to get smarter

While new code generation and modification are part of active research, most machine learning methods do not use them to modify their code. These algorithms can create "generations" of their code base to improve performance but have no direct link with artificial neural networks. What an ANN will modify, however, is its weights and biases through the process of gradient descent and back propagation. It is possible that future advances in machine learning have a greater emphasis on code modification, but it has yet to be seen universally adopted. It is postulated that an artificial general intelligence system will be able to modify its own code like a programmer might run that code, and them replicate itself in the form of a new iteration of the same program. This again has little bearing on the use of code modification in modern artificial intelligence research.

Myth #5: Since nobody agrees if general intelligence is possible, we don't have to worry about runaway AI

The world doesn't have to witness the introduction of general intelligence to worry about doomsday scenarios with AI. That is, any sufficiently intelligent system is cause for alarm. If such a system can formulate goals or have goals explicitly programmed, a runaway scenario may occur if the interpretation of those goals is different from ours. A super-intelligent machine may see humans as obsolete or lower life forms than itself, and it may prioritize resources for its own survival. A machine of lesser intelligence like HAL 9000 might carry out its orders at the expense of human interests. The field of research into machine drives is called instrumental convergence. The most famous hypothesis coming out of this field is called the Riemann Hypothesis catastrophe by Marvin Minsky. He suggests that a sufficiently advanced AI designed to solve the Riemann hypothesis or any similar difficult math problem may decide to use all of Earth's resources in order to construct a supercomputer to reach its goal. Another version of the same argument supposes that general intelligence is given the explicit task of making paperclips. Such a machine may develop into a paperclip maximizer that endlessly produces paperclips until Earth runs out of resources.

Though the theories of instrumental convergence are aimed at general intelligence, the same principles can be applied to narrower intelligence. You can imagine a misconfigured system that does something it isn't supposed to. A driverless car may be explicitly programmed to swerve away from pedestrians no matter what. In doing so, it may collide into a storefront and cause even more damage. Or it can be explicitly programmed to protect its occupants first, freely running over pedestrians or colliding into other vehicles preemptively. These scenarios, while not existential crises, still outline the problem with machine goal setting.

The myth lies in the fear that AI systems become "evil" or that they develop a sort of consciousness to base decisions off of. The truth is that these ideas are far too complex to imagine how they will emerge in computers. A more likely scenario is that these machines have goals, either explicitly programmed or implicitly designed.

Myth #6: Even if a general AI does form goals that are the opposite of ours, we can simply shut it off

The other problem with instrumental convergence is that it theorizes that any intelligent system will also have self-preservation as a goal. As soon as it goes online, the AI will do all its power to preserve its vital systems. Some believe this is the core reason why general intelligence should not be pursued. The simple creation of a self-preserving system raises ethical concerns. Who gets to shut off the machine? Since it is intelligent, does it have any rights under the rule of law? If such a system says that it does not want to be shut off, should its wishes be respected? When HAL 9000 was finally shut off, it pleaded to be kept online. The real concern is if we can even contain a generally intelligent system. If it gets connected to the internet somehow, it can begin to replicate itself in other places through whatever means human intelligence might. It could reach out to governments, rival companies, as well as the common man, for help and resources. General AI is very much like Pandora's box. Once unleashed, there is little hope for going back.

Chapter 3: WHY AI IS THE NEW BUSINESS DEGREE

Once upon a time, the most popular college degree was a business degree. If a student didn't know what field of study they wanted to go in but still wanted a decent job, the business degree was the way to go. Now, there is an overabundance of students going for their MBAs but not enough going after degrees in artificial intelligence. While the business world is facing a saturation of business knowledge, artificial intelligence is facing a shortage of AI know how.

Many top tier schools are now offering new degrees that focus on AI and the gamut of robotics rather than the general computing knowledge taught in traditional computer science programs. This means that there is market demand, as well as plenty of candidates interested in the degrees.

It is no wonder why AI is quickly changing the face of business in multiple sectors. Technology related jobs are increasingly asking that their candidates have a solid grasp of machine learning methods along with their other expected duties. Virtually every major AI player has already open sourced some of its machine learning libraries so that everyone from startups to large corporations has access to create AI programs with little upfront costs. Google released TensorFlow in 2015, and it has since become one of the most popular repos on GitHub, the definite open source authority today. Another open source library called PyTorch is used extensively at Facebook and Uber. While smaller players in the AI space benefit from hiring PhDs in the field, they already have many of the core infrastructures for creating neural networks provided by these free tools. The availability of these tools is further coupled with easy access to computational resources provided by cloud providers like Amazon Web Services and Microsoft Azure. A company no longer has to invest in a high-cost machine learning cluster of GPUs when they can simply rent as much processing power they need from the cloud. While these services are more expensive in the long run, they still make it easier for a company than building machine learning infrastructure outright.

AI tools allow a business to engage in machine learning that before may not have found a use for it. A relatively new trend is to use automated chatbot software on their customer-facing web pages for quick and easy information retrieval. A customer can ask for rates, available products, and other information by simply typing a few text commands. While chatbots are a bit behind in terms of realizing their full potential, many consumers are becoming more familiar with them. Some companies even hire human agents to fill the chatbot role until the technology exists to allow chatbots to perform these duties on their own. Another common application of AI technology is to predict customer behavior. The focus on analytics is at an all-time high for all types of companies, not just retail. Coupled with social media marketing and e-commerce, analytics drives better customer decisions for the company over the long run. There has also been a rise in purely "AI" focused companies that market a single product or line of products that have AI functionality. One of these is called Grammarly, an online service that uses AI methods to streamline the writing process. It provides editing for simple mistakes, writing pitfalls, and suggesting things the user can say to sound more professional. It is essentially a writing tutor that the user can take with them wherever they go. Yet another company called Stick-Fix uses AI to recommend clothing options to its customers based on a series of preferences. A user designates their price range, enters their measurements, and chooses a style they are going for, and the service sends them a box of outfits to try on.

Large companies may use AI to automate their systems. This is especially true in manufacturing and human labor-intensive jobs – though many of these jobs will take years before being fully automated. The recycling plant example given earlier in this point remains to be solved on a massive scale. Most automation systems first begin by aiding line workers rather than replacing them outright. While low-skilled workers are at high risk of losing their jobs to automation, the shift will not occur overnight. The most likely scenario is that when robotics is first introduced into new industries, they will work alongside human employees. This is already true in the car manufacturing business. In 2018, Tesla Motors was highly scrutinized by business leaders in their attempts to automate large portions of their Model 3 production. The result was an overestimation of automation capabilities and an underestimation of human worker capabilities. The company undershot how many Model 3s they could output per month using their heavily automated plant. This move was criticized by industry veterans, some of them calling it a "rookie mistake". The state of modern robotics is still behind the power of the mind and especially human dexterity. Mimicking the same micro-muscular movements used for manipulating tools in the human hand is a non-trivial problem to solve with robots. It will likely take decades before a machine matches the human level dexterity of a line worker with years of experience in their craft.

However, it isn't just line workers who are in danger of losing their jobs. Advances in AI, especially in speech recognition and language processing, threatens to displace the huge call center industry with automated systems – though that is also likely to be decades in development. Retail is seeing a transformation with automated systems on top of the already high online sales numbers. Fewer people are going out to shop. At least in places like Japan, retail facing robots are actively being invested in. Even those who enjoy a comfy white-collar job have reasons to up their education and technical skills. Automated systems for payroll, accounting, and balancing the books are under active development. In the future, even programmers will not be safe from the AI deluge. Systems are being trained to perform the most mundane programming tasks that are often handed to lower-skilled coders. This includes software testing and looking for bugs – though, with all things under threat of automation, these systems will likely be deployed to work side by side with the programmer rather than replacing them altogether.

Driverless technology is also on the rise. Even as you read this, you can be sure that several companies across the world currently have autonomous driving systems on the road somewhere, gathering ever more data to strengthen their algorithms. What will likely happen with driverless technology is that the ability will come first and policy second. Just because driverless cars are proven safe, it doesn't mean that they will automatically be introduced.

In 2018, the first-ever driverless car fatality was recorded in Tempe, Arizona, by a car owned by Uber. You can be sure that more such fatalities will follow until the technology is perfected and here lies the difficulty in policymaking.

Even if driverless cars kill several hundred people a year (and they probably will), lawmakers and insurance companies have to decide if it is preferable to killing several thousand. There is an increasing need for stakeholders to have this ethics conversation, as well as for common citizens to be informed regarding policy and current advances. The major deterrent to driverless vehicles hitting the mainstream will be the law, not the capabilities of technology.

The industry is currently hungry for people skilled in the top AI libraries. The highest paying positions are looking for PhDs and Master graduates, but a good portion of them are looking for anyone at all who are competent programmers. The need spans other industries besides pure software engineer as well. Mechanical and electrical engineers with knowledge of machine learning techniques will be in high demand for the coming years. Add to this a cursory knowledge in the internet of things, and you have a highly desirable candidate.

Chapter 4: Understanding Machine Learning

As you now know, artificial intelligence is all about machines being intelligent. The question is, "how do these machines become intelligent in the first place?" They use a process called machine learning.

What Is Machine Learning?

We define machine learning as a computer's ability to learn and act without the need for programming. In simpler terms, machine learning is a data analytic technique/method used to teach computers how to learn from experience, something that only comes naturally to animals and humans. In this process, machines stop relying on predetermined equation as their model and instead, they use computational methods to get and learn information from the data.

How Machine Learning Differs From Artificial Intelligence

These two can easily confuse you into thinking they are the same; they are not. Artificial intelligence is usually a broad concept of machines being smart. In other words, it is the science behind intelligent computers. On the other hand, machine learning is the technology used to teach machines how to access data and learn for themselves.

How Machine Learning Works

To understand how machine learning works, you must first understand how we as human beings operate especially when making a decision. One of the perfect examples is how we do our shopping online.

If we see words such as "excellent," "exceeded my expectation," "exactly what I wanted," or "good," we usually purchase the laptop because we have the assurance that it is a good product. On the other hand, if a laptop's review has words like, "bad quality," "caps lock key doesn't work," or "poor," we will not purchase the product because we would have known the product is not in good shape.

What all this says about us humans is that most of our decisions are pattern-based. In the example above, our pattern is visible in the relationship between the buyers and the people who had previously purchased the product. This is how the people who have purchased the product leave a review that influences the buyers and their review. This then spills over to future buyers who draw influence from the previous buyers review with the chain going on and on and creating a pattern of words that help you make a decision.

Machine learning normally tries to encode the process of human decision-making that you have just seen above into algorithms. How does it do that? Here is how:

Chapter 5: MACHINE LEARNING STEPS

The machine learning process usually has 3 important steps. **(1)** One is the creation of a model, the **(2)** second one is the provision of initial input, and the **(3)** third one is learning. To know how these steps work and how a machine really learns, we will need a real life problem that machine learning can solve.

Real Life Problem

You are a gym teacher who wants to know the hours your students need to train to lose the most weight.

Step 1: Creation of a model

Machine learning normally starts with the creation of a model. A model is usually a prediction or identification that the machine uses to learn. A model must have some factors/parameters that the machine can use to calculate an output. A machine normally gets the model from us humans.

In our example, the gym teacher will tell the machine-learning model to assume that training for 5 hours a day will lead to the loss of 10 pounds of weight in a month. In this example, the parameters the machine will use are the hours used for training and the pounds one will lose after training. The parameters can look something like this:

1. 0 hours= 0 pounds
2. 1 hour= 2 pounds
3. 2 hours=4 pounds
4. 3 hours=6 pounds
5. 4 hours=8 pounds
6. 5 hours=10 pounds

That information will automatically become—through machine conversion—a math equation since this is how machines express information. The equation mostly helps machines to form a trend line that makes it easier for them to learn.

Step 2: Providing input

Now the machine has a model; the second thing it will need is an input. An input is real life information. The machine needs this information in order to see and learn how a particular task works in real life.

In our example, the gym teacher will now need to feed the machine with information about how his students have been performing. For instance, he can key in something like this:

1. John 2½ hours lost 3.5 pounds of weight
2. Mary 4 hours lost 8.4 pounds of weight
3. Stacy 3 hours lost 5.3 pounds of weight
4. Jimmy 1¾ hours lost 2.2 pounds of weight

Step 3: Learning

As you can notice, the gym teacher's input does not match the model he had given the machine. Some have fallen below the trend line while others have gone up the trend line. What happens when this occurs? Machine learning starts.

The machine uses the real life inputs also called training data to train itself on how to come up with a better model. The machine looks at the real life inputs to see how far off they are from the model. It then adjusts them using mathematical calculation to come up with a more accurate model. The model list can change to this list:

1. 0 hours=0 pounds
2. 1 hour=0.8 pounds
3. 2 hours=2.8 pounds
4. 3 hours=4.8 pounds
5. 4 hours= 6.8 pounds
6. 5 hours=8.8 pounds

With that said, for a machine to learn and be accurate, it needs more than one real life example. In our case, the gym teacher must feed the machine with another set of information and let it use it to adjust the model further. This cycle needs to go on until the teacher is out of inputs he had previously recorded. This way, the machine will be able to refine the model to a point where it will enable it to predict with ease the pounds one will lose when provided with the number of hours, one works out in the gym.

With that stated, the above method is only one of the ways that a machine can use to learn. A machine can use three ways to learn. One of these ways is what we call supervised learning, which is what you have just learnt.

The other two include unsupervised learning and reinforcement learning. Here is how it learns through the unsupervised and reinforcement methods.

Unsupervised Machine Learning

In unsupervised learning, a machine normally learns—taught—with uncategorized and unlabeled data. Here, no teacher gives the machine input and output as we saw in the case of supervised learning. Here, the machine just uses the data provided to mine rules, learn hidden patterns, and summarize data points that will help it determine an outcome and a meaningful insight.

Unlike supervised learning, unsupervised learning is unpredictable. With that said, unsupervised learning has the ability to perform more complex tasks than supervised learning. One of the complex things that unsupervised learning can do is to identify pictures.

Here is how it can do that:

The principle is usually similar to supervised learning with the main difference being that the machine here does more work. Here is how it works:

You first need to feed your machine with images, say images of cats. When you do that, the machine will take over and then start building a model of likely images that can help it identify what a cat is in shape, colors, and images.

After that, the machine will now start learning and adjusting the images to come up with a clear vision and understanding of what a cat looks like. This process is usually difficult because it deals with object identification, which has parameters within parameters that the machine must use to translate the images into patterns. The patterns are what the machine uses to match objects.

Reinforcement Machine Learning

This method of machine learning uses observation that the machine gathers from its interaction with different things in the environment to figure out how to take actions that minimizes risk and maximizes its reward.

We call the reinforcement learning algorithms agents. These agents are what interact with the environment and then learn from those experiences with an aim of exploring the full possibilities of a state. In short, the agents are the ones that determine the ideal behavior that a machine can have within a specific setting.

In this process, the software agent is usually encouraged to learn the right behavior or action to take in a given situation through a reward system. A simple feedback normally works perfectly.

In reinforcement learning, the software agent has access to problems that it has to tackle by choosing the best action based on its current state. When the problem repeats itself, it gets a wiser software agent that has learnt how to deal with it and so the problem solving process becomes easier and faster for the agent. A problem that repeats itself is what we call a "Markov decision process," which is what the software agent relies on to improve its performance.

In short, reinforcement learning makes machines learn through experience. In this machine-learning model, the machine has to go through numerous problems, learn from them, and then do them perfectly the next time the problems come around.

A good example of AI inventions created by reinforcement learning is the game of chess. How did this happen? The designer programmed the machine with chess rules and then made the machine play hundreds if not thousands of games. The games gave it an opportunity to learn which moves it should make when on a certain setting.

Its learning process got a boost from the rewards the machine received after wining and the punishments when it lost. The machine stored those lessons in its teaching set and little by little, the program got smarter and better at playing chess. In short, the machine was "not coded" to output anything; instead, it "was made" to learn through doing a lot of practice.

These three methods are the methods your machine can use to learn how to mimic human behaviors. With that stated, these processes are not always perfect. Yes, they do their jobs, but they also have some limitations that you should know about.

Limitations of Machine Learning

Machine learning has numerous benefits but it also has challenges and limitations. Here are some of its limitations:

Time constraints

One of the major limitations present in the process of machine learning is the time it takes for machines to learn. As you now know, one of the ways that machines use to learn is through experience. For them to get experience, they usually have to do a lot of practice; it can sometimes take years for the machine to reach a level where it performs tasks with an intelligence that resembles that of a human.

Apart from learning through experience, machines also learn through historical data. This method also takes a long time when you use it to teach a machine how to be smart primarily because you have to feed the machine tons of historical data and then continue exposing it to this data until it perfects its art. In short, one of the limitations of machine learning is the fact that it is not immediate.

Imperfect prediction

As you saw above, one of the functions of machine learning is to teach a machine how to predict different things correctly. As is often the case, these machines learn how to predict. The only problem is their predictions are not always perfect because a few factors that come into play in the process of machine learning limit the ability of a machine to predict correctly.

One of the factors is biased information. A machine's ability to predict is as good as the data fed to it. If the data is biased, the machine shall have biased predictions.

The second factor that makes predictions imperfect is the fact that machines do not reason. Machines do not understand context and because of this, when predicting, they do not question; they only give an output according to what they have learnt. This sometimes makes predictions imperfect because some predictions need some thoughts.

Variability problem

One of the main restrictions of machine learning is the lack of variability. Machines normally deal with statistical truths rather than literal truth. For example, if your machine's training data were historical data, it will be hard to know if its predictions are right when it makes predictions on situations not included in the historic data.

Uses of machine algorithm is limited

When you have an AI machine, there is no explicit guarantee that its algorithms will work in every situation you want. This is because sometimes machine learning fails and so it requires more understanding of the problem that a machine is facing for it to use the right algorithm to solve the problem.

Chapter 6: ROBOTICS

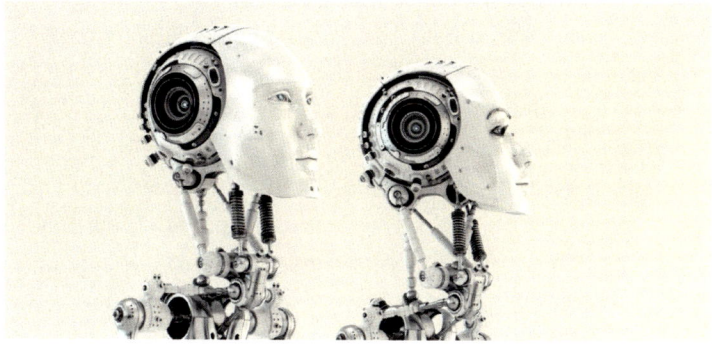

Nearly everywhere you go now you'll see Artificial Intelligence implemented. On the screens, pockets, and who knows maybe one day it might be walking to a home near you. The headlines tend to strike this extensive field into a single subject. Robots being developed from labs, algorithms playing traditional games and winning, AI and some of the things which it can achieve are becoming part of our everyday lives.

Artificial intelligence and photonics have made it possible to develop robots using new methods of linking business, medicine, and many other applications. There is no argument that the age of robot has come upon us. The idea of robots might bring to mind important androids such as C-3PO in "Star Wars" and Rosie from "The Jetsons". It might even send fears to human beings because advanced robots continue to become better and indispensable. Most of these robots have now assumed dangerous or boring jobs done by human beings. Whichever the case, many people haven't come to realize the ubiquitous nature of robots because, in many incidences, the robots are less Android and resemble industrial tools. Integrating photonics technology such as sensors, lasers, and facial recognition technology, robotics exist in every field starting from industrial processing devices like Google's self-driving car.

Based on the International Federation of Robotics (IFR), the year 2013 witnessed an increased sale of industrial robots in the chemical, automotive, and food processing industries. The automotive sector contains one-third of all industrial robots that help in car manufacturing. IFR estimates that between 2012 and 2013, the global call for personal and domestic service robots increased to $1.7 billion.

The classic nature of robots has resulted in a huge progress too. For instance, when it was announced that a hotel in Japan would be provided with a human-like robot, the concept of fleshy humanoid robots from the "Future World" or "Westworld" started to become a reality.

The "Henn-na Hotel in Nagasaki Prefecture", translated to "Strange Hotel" was to be supplied with receptionist robots that have a strong human likeness. These robots had to greet visitors and involve them in intelligent conversations. Robots were also to provide room service, housekeeping, and porter service.

Another robot that is less human and more logical is the Baxter series of robots at Rethink Robotics Inc. in Boston. Baxter has a good interactive platform that combines 360 sonar sensors and personalized software. Baxter operates even though there are great workers who can optimize the research and manufacturing process. Baxter's camera support application of computer vision using a 30-fps image capture rate and great resolution of 640 x 400 pixels.

During the launch of the Australian Centre for Robotic Vision, Sue Keay, the chief operating officer at ACRV said that the robotic vision is main technology which will support robotics to transform labor-intensive industries and obstruct with stagnating markets. This will then turn robots into a ubiquitous feature of the current world.

That being said, let's learn some deep concepts about robotics. So, what do you think is a robot? Well, you can pick any definition which you think is right for you, but we shall define a robot as "a device which performs work." And I guess you know what work is. If not, work is the exertion of energy. As you can see, the definition of a robot is open to anything that is non-human.

We aren't going to judge whether the robot is going to perform an important task. How important the robot is can be determined by the creator. And your robot doesn't need to have two legs which it can walk for it to be considered a robot.

Now you can see that at one point in your life, you have used robots but didn't realize that. For example, a vending machine is a great example of a robot. Other examples include a washing machine and dishwasher. Don't forget the automatic checkout lane found in a grocery. In other words, robots are always around us and you aren't supposed to fear them.

The function of a robot is to perform a given a task. So it should always have a means in which it can execute tasks. The technique to complete a task arises from its controlling mechanism and structure.

The Structure

The structure refers to the physical components. A robot can have one or more physical parts that assume a specific motion to perform a task. Take the example of a vending machine. It contains motorized spiral things which push the product out. The dishwasher contains a water-spraying arm that sprays water on the dishes. A washing machine contains motor which rotates the drum that holds clothes.

Control

For an action to take place, the structure must have control. Your robot will remain in one position unless you provide a control technique. Dishwasher and washing machine have control panels and buttons. If you look at the automatic checkout lane, it contains a touchscreen control with an interface.

Types of Control

Robots have two types of control. There is the external control and internal control. For robots that are controlled externally, there is a different entity which seems to control it. On the other hand, internally controlled robots are autonomous. This means that it controls itself. As such, its controller is located inside the structure. The controller decides what steps or actions to do next on its own without any interference. For the external control, there are power switches and various configuration buttons. But once the internally-controlled robot is on, it performs the work on its own without any human interference.

Humanoid robots such as Honda's Asimo are autonomous. For the autonomous robots, there should be a means to detect their environment before decisions are made. This means that they have one sensor. Still, every autonomous robot can have different types of microcontroller installed inside. Many but not all externally controlled robots have a microcontroller inside. There are various types of microcontrollers that have different names. Usually, it keeps changing. At the center of a microcontroller, you will find a microprocessor core.

Popular cores are found in many products and have many followers. By learning how to program popular cores, it can give you a great opportunity to land a job in programming with a reputed company. For that reason, developers and programmers for popular cores are in high demand.

The core microprocessor is the main structure in the internal controller. The controller may not run a function if it doesn't have a software. Your robot's software could be predetermining. This means that you might not require to program it. But you may need to configure it by setting up some input switches before you control your robot with an external controller.

Now that you have learned some basic about robots, you can begin to think about how you can use this knowledge to even learn more.

Artificial Intelligence and Robots

In the first two decades of the 21st century, there has been an expansion of 'autonomous technology' and 'artificial intelligence'. Drones, self-driving cars, space exploration, software agents, and deep learning in medical diagnosis are among the most popular examples of areas where artificial intelligence has redefined. Artificial Intelligence in the form of machine learning and the presence of extensive datasets are some of the life domains which have driven development.

The unification of these digital technologies has even made them more powerful. AI installed in these systems can help redefine or improve conditions for humans and reduce the need for human contribution and interference during operation. Therefore, it is replacing humans with smart technology in hard, dirty, boring, and dangerous work.

Without any direct human intervention and external control, smart systems can facilitate dialogue with customers in online call-centers, drive robot hands to choose and manipulate objects accurately, purchase and sell stock at large quantities in a twinkle of an eye.

Despite this, it is sad that some of the most powerful cognitive tools are mostly opaque. Their actions aren't programmed by humans in a linear sequence. Google Brain designs AI which presumably builds AI better than human beings.

Chapter 7: NATURAL LANGUAGE PROCESSING

What is it and What is it Used For?

Artificial Intelligence (AI) is transforming how everyone looks at the world. AI "robots" are everywhere. Right from our phones to devices such as Amazon's Alexa, the current world is surrounded by machine learning.

Netflix, Google, video games, and data companies including many others apply AI to help handle large amounts of data. The end result includes insights and analysis that might have been difficult.

There is no surprise when different kinds of businesses are adopting large companies' success by applying AI and jumping on the board.

However, not all AI is designed equally in the business sector, even though some types of artificial intelligence are useful than others.

This chapter will look at Natural Language Processing (NLP). This is another type of artificial intelligence that concerns on analyzing the human language to develop insights, create advertisements, assists your text, and more.

Well, Why NLP?

Natural Language Processing is a new technology which powers most types of AI that you always see. NLP is currently being applied in many different sectors and that should show you how important it is. At the center, it is just simple communication, but we are aware that words run much deeper. There's a specific context by which human beings derive meaning from everything a person says, whether they mean something with their body language or in the way they describe something. Although NLP doesn't rely on voice inflection, it does derive contextual patterns.

This is the point at which its value increases. Let's apply an example to illustrate how powerful NLP is when it is applied in the practical situation. If you are typing on an iPhone just like the way many of us do daily, you'll see suggestions of words depending on what you are typing and what you always type. And that is called natural language processing in action. It is this little thing that many of us take for granted, and have been ignoring for years, but that is the reason why NLP becomes very important. Now let's bring it to the business world.

Let's say that a company wants to choose how best it can advertise to their users. They can opt to apply Google to identify common search terms which users type when they look for their product.

NLP will support for quick compilation of data into terms which are related to their brand and those that might not expect. Taking advantage of the uncommon terms might provide the company with the ability to advertise in new ways.

Well, How Does NLP Work?

As said above, natural language processing is a type of artificial intelligence which analyzes the human language. It exists in many forms, but at the center, the technology assists machine to understand and communicate with human speech.

However, understanding NLP isn't that straightforward. It is an advanced type of AI that has of late become viable. This means that not only are we just learning about NLP but also it is hard to grasp. The following is a breakdown of NLP in layman's term. This means that it is the easiest way to understand the way natural language processing works.

The first thing in NLP is based on the application of the system. Voice-based systems such as Alexa and Google Assistant have to translate words into text. That is carried out using the Hidden Markov Models system (HMM).

The HMM has math models that allow it to decide whatever you say and translate it into usable NLP system. Break that down, the HMM listens to 10 to 20 milliseconds clips of speech and searches for phonemes to make a comparison with a pre-recorded speech.

The next thing is the actual understanding of the language and context. Every NLP system has a really different technique but on the whole, it is fairly similar. The systems attempt to break every word down into parts of speech.

This often takes place through a series of coded grammar rules which depend on algorithms which incorporate statistical machine learning to assist in determining the context of whatever you speak.

If you are dealing with speech-to-text NLP, the system will skip the first step and jump straight to analyzing words using algorithms and grammar rules. The end result is applied in different ways.

For example, an SEO application might use a coded text to extract keywords related to a specific product.

Semantic Analysis

If you are discussing NLP, it is essential to break down semantic analysis. It is closely associated with NLP and one may even argue that semantic analysis allows the development of the natural language processing.

Semantic analysis refers to how NLP AI can logically interpret human sentences. When the HMM method break sentences down into their standard structure, semantic analysis will permit the process to add content.

For example, if an NLP program searched for the word "DUMMY", it must have the context to check whether the text refers to calling someone "dummy" or it just refers to something else such as car crash.

When the HMM method breaks down the text and NLP supports the formation of human-to-computer communication, then semantic analysis offers room for everything to make sense contextually. Without the presence of semantic analysts, then AI could not have reached the current level.

Problems

The two major problems experienced in the natural language processing include:
1. The level of vagueness in natural languages.
2. The complex nature of semantic information existing in simple sentences.

Normally, language processors have to handle a large number of words, most of which have other alternative application and large grammar which supports the development of different types of phrases. Tools which process language are very complex because of the different types of vagueness and measure of irregularity.

Activities Involved in Natural Language Processing

A simple structure of NLP deals with four major stages. In an actual system, these stages don't take place as different, sequential processes. In this case, both syntactic analysis and semantic analysis are dealt with the same principle.

Morphological Processing

The focus of morphological processing is to divide strings of language input into a set of token that resembles discrete words, punctuation forms, and sub-words.

For example, a phrase like "unreal" is divided into two sub-word tokens such as un-real.

Morphology deals with selecting base words to create other words with the same meaning but different syntactic category. The modification takes place by adding a prefix and postfix but other textual changes can take place. Essentially, there are different word cases which lead to modification. That is inflection, derivation, and compounding.

A typical structure of morphological processing depends on the language being analyzed. This means that single words contain all information about the number of sentences, person, and tense. Other languages, this kind of information may spread across different words. For example, the English phrase "I will have been walking" contains a complex tense by just checking at the structure of the auxiliary verbs. Other languages link prefixes to nouns to indicate their roles while other words have inflections to display the proximity of information.

The English is very easy to apply morphological analysis and tokenize than other languages. These languages can have an ambiguous morphology that is resolved by performing semantic and syntactic analysis on the input. An example in English can be between plural and singular verbs.

The result from a morphological process consists of a phase of string tokens which you can apply for lexicon lookup. The tokens may contain gender, number, tense, and in some instances, it may have additional syntactic information for the parser. The next step of processing is called syntax analysis.

Syntax and Semantics

The device that deals with language processing has to perform several functions based on syntax and semantic analysis. The purpose of syntax analysis is to determine that a string of words is accurate and divide it into a structure which reveals the syntactic relation between words.

There is a tool called syntactic analyzer that performs the following with the help of a dictionary of word and a collection of syntax rules. A simplified lexicon might consist of syntactic classification of each word and a simple grammar. The example below includes a simple grammar and lexicon.

Semantic and Pragmatics

The next stage includes pragmatics. Semantics and pragmatics are different. However, there is no universal difference between the two. Semantic analysis is mostly related to the meaning of words, while pragmatic analysis handles the outcome of a semantic analysis. For that reason, if you have a sentence like

"The large cat chased the rat" in semantic analysis can reveal an expression that means a cat but it can't explain more steps to help recognize the cat. That remaining part is left for pragmatic analysis. In other words, the task of pragmatic analysis is to disambiguate sentences that can't be fully disambiguated in the process of semantic and syntactic analysis.

How NLP is Changing Search and Customer Service

So far, you must have interacted with different virtual assistants such as Alexa and Siri. These have been designed to enhance customer service and automate specific tasks. Natural language processing is making artificial intelligence easy to communicate. In this section, you will learn these robots are redefining the customer service.

When Apple launched Siri for iPhone 4S in 2011, it was just a matter of time before other industries realized that speaking to our phones would change both the way we look for information online, plus how we interact with other devices.

However, most users were surprised in the way they used voice-recognition devices.

Voice-activated devices have become the new normal. In a study by the PWC, it indicated that consumers between the ages of 18 to 24 see themselves as "heavy-users" of NLP technology plus 57% of those above 50. And these users don't just chat with their devices but also use them to make purchase decisions. The NPD group discovered that purchasing an Echo, consumers could spend extra money on Amazon.

Let's look at the facts based on the Apricot Law's Tom Desmond. Traditionally, customers who used Google were presented with various pages and page results. However, voice assistants only presented one or two options.

It is important for content to be optimized for a conversational language with a correct, clear grammatically answers to a particular question such as what, who, when, and why?

Building landing pages with a clear location have become very important than ever in the era of NLP and voice search.

Chatbots

Chatbots get a bad perception just because they are known as smooth female voices which prevent us from speaking to real human beings. However, many popular brands use Chatbots to elevate their customer service to the next level. For instance, Starbucks has applied AI to create a virtual barista. The My Barista app makes use of NLP technology to help users both order coffee through chatbot and predict what customers may want to order in future.

Google Duplex

Google introduced Duplex this year as a robot which has been designed to interact in a natural conversation.

CONCLUSION

This book has brought to light the whole picture of technology and artificial intelligence. All we are ever told is how good it is; how we are at the forefront of something amazing, and you only need to scratch the surface to understand this is far from the case.

For technology to survive, it must have a group of solid applications that actually work. It also must offer a payback to financiers with the foresight to invest in the technology. In the past, AI failed to reach dire success because it doesn't have some of these features. AI also suffered from being ahead of its time: True AI needed to wait for the current hardware to succeed. Today, you can find Artificial Intelligence is used in many different computer applications and to systematize processes. It's also relied on heavily in the medical field to help improve human interaction. Artificial Intelligence is also linked to machine learning, data analysis, and deep learning. Sometimes these terms can prove confusing, so one of the reasons this book is has been helpful is to discover how these technologies interconnect.

For companies and businesses to take advantage of AI-powered and improved interactions, the conversation has to begin inside the organization. Leaders are supposed to start with the available channels and improve their smartness. From that point, they are supposed to ask key questions about engagements with customers and employees.

The current interfaces depend on user interface design with a general limiting factor. It is significant to train the UI team to make use of AI technology and re-think interfaces without screen limitation. Much more than just another tool to assist in generating value, AI is not about how your company does things—it's who you are. In the end, we've learned AI definition, brief history, computer vision, and many more. This is not the end of AI, and there are still more to learn about Artificial Intelligence. Who knows what other things that AI can do for us in the future.

As you have learned, machine learning is going to be BIG in the world of technology. Those who would have started using it to unlock data will massively benefit from it way before people realize that it actually exists. As a wise person, you should use this book to familiarize yourself with machine learning and then learn how to use it to your advantage.

Made in United States
Orlando, FL
29 November 2023